Minutes a Day-Mastery for a Lifetime!

Level 2

Mathematics
2nd Edition

1st semester

Nancy L. McGraw

Bright Ideas Press, LLC
Cleveland, Ohio

Simple Solutions Level 2 Second Edition

1ˢᵗ semester

All rights reserved. No part of this publication may be reproduced or transmitted in any form or by any means, electronic or mechanical, including photocopy, recording, or any information storage or retrieval system. Reproduction of these materials for an entire class, school, or district is prohibited.

Printed in the United States of America

ISBN-13: 978-1-934210-31-4
ISBN-10: 1-934210-31-5

Cover Design: Dan Mazzola
Editor: Kimberly A. Dambrogio

Copyright © 2009 by Bright Ideas Press, LLC
Cleveland, Ohio

Welcome to Simple Solutions

Note to the Student:

This workbook will give you the opportunity to practice skills you have learned in previous grades. By practicing these skills each day, you will gain confidence in your math ability.

Using this workbook will help you understand math concepts more easily and for many of you, will give you a more positive attitude towards math in general.

In order for this program to help you be successful, it is extremely important that you do a lesson every day. It is also important that you check your answers and ask your teacher for help with the problems you didn't understand or that you did incorrectly.

If you put forth the effort, Simple Solutions will change your opinion about math forever.

Simple Solutions® Mathematics — Level 2, 1st semester

Lesson #1

1. Write the word for the number of circles in each group.

   a.

   b.

   c.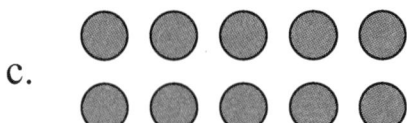

2. Which number is greater, *25* or *16*?

3. 3 + 2 = ?

4. 6 – 1 = ?

5. What should come next in the pattern? ○ ○ ◯ ○ ○

6. 0 + 5 = ?

7. Write the number that goes between *39* and *41*.   39 ____ 41

8. What number would come next?   2, 4, 6, ____

9. How much money is 1 nickel and 3 pennies?

10. 7 – 7 = ?

11. What number is missing from the list?   30, ____, 50, 60, 70

12. 10 – 7 = ?

13. What is the value of a quarter?   45¢   25¢   30¢

14. What is the name of this shape? ◯

15. There are 6 cats and 4 dogs at the pet shop. How many more cats are at the shop than dogs?

Simple Solutions© Mathematics — Level 2, 1st semester

1.	2.	3.
4.	5.	6.
7.	8.	9.
10.	11.	12.
13.	14.	15.

Simple Solutions© Mathematics   Level 2, 1st semester

Lesson #2

1. There are 3 cows and 4 sheep on a farm. How many animals are there altogether? Did you add or subtract to solve this problem?

2. $12 - 6 = ?$

3. Five pennies are the same as 1 _____.   dime   nickel   quarter

4. **The symbol (>) means *greater than*. The symbol (<) means *less than*. The open part of the sign points toward the bigger number.** Put the correct symbol in the circle.   13 ◯ 21

5. $3 + 8 = ?$

6. Name this shape.

7. $9 - 2 = ?$

8. What number will come next?   10, 20, 30, ____

9. I have 6 tens and 2 ones. Which number am I?   32   26   62

10. Color $\frac{1}{2}$ of the shape.

11. $4 + 4 = ?$

12. What number comes after *33*?   33, ____

13. A box of spaghetti weighs about 1 pound. Would a toaster weigh more than a pound or less than a pound?

14. Write the time shown on this clock.

15. What number comes between *43* and *45*?   43 ____ 45

1.	2.	3.
4.	5.	6.
7.	8.	9.
10.	11.	12.
13.	14.	15.

Lesson #3

1. Does it take about 1 minute or 1 hour to brush your teeth?

2. Put the numbers in order from smallest to largest.   35, 31, 36

3. 3 + 4 = ?

4. The number *23* has 2 tens and 3 ones. What number has 5 tens and 7 ones?

5. Name the shape.

6. How many parts out of 4 parts are shaded?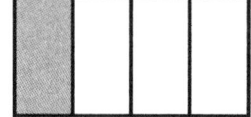

7. 10 − 4 = ?

8. What time is shown on the clock?

9. 2 + 6 = ?

10. Sarah has 2 dimes and 4 pennies. How much money does she have?

11. What number comes right before *78*?   ____, 78

12. Jared found 8 shells. He gave 5 shells to Mike. How many shells does he have left?

13. 12 − 7 = ?

14. What is the name of the month?

    What is the date of the first Monday of the month?

15. How many days are in one week?

April

S	M	T	W	T	F	S
				1	2	3
4	5	6	7	8	9	10
11	12	13	14	15	16	17
18	19	20	21	22	23	24
25	26	27	28	29	30	

Simple Solutions® Mathematics — Level 2, 1st semester

1.	2.	3.
4.	5.	6.
7.	8.	9.
10.	11.	12.
13.	14.	15.

Lesson #4

1. What number would come next?   5, 10, 15, _____

2. 7 + 7 = ?

3. What is the time on the clock?

4. What number comes between *68* and *70*?   68 _____ 70

5. Put the correct symbol (< or >) in the circle.   27 ◯ 42

6. 9 – 3 = ?

7. Name the shape.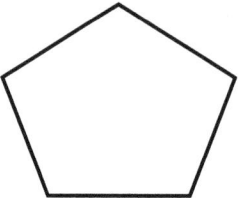

8. Marie has 2 dimes. How much money does she have?

9. *73* has 7 tens and _____ ones.

10. Shawn had 4 pencils. Danielle gave him 5 more pencils. How many pencils does Shawn have now?

11. 2 + 8 = ?

12. Shade $\frac{1}{2}$ of the shape.

13. 13 – 8 = ?

14. Is the number *21* greater than or less than *29*?

15. **Use the table.** How many picture books are in the classroom library?

    How many picture books and joke books are there in all?

Classroom Library	
Number	Book Type
15	Science
10	Picture
3	Joke

1.	2.	3.
4.	5.	6.
7.	8.	9.
10.	11.	12.
13.	14.	15.

Lesson #5

1. The number *69* has ____ tens and ____ ones.

2. Put the correct symbol (< or >) in the circle. 46 ◯ 32

3. Write the time that is shown on the clock.

4. 9 + 8 = ?

5. What number comes right after *79*?   79, ____

6. 4 − 2 = ?

7. What is the name of this shape?

8. Write the word for the number of circles in the group.

9. Is a door taller or shorter than you are?

10. 16 − 8 = ?

11. There are 7 bluebirds in one tree and 3 robins in another tree. How many birds are there in all?

12. 3 + 7 + 2 = ?

13. What number comes next?   11, 12, 13, ____

14. How many parts are shaded out of 3 parts?

15. A nickel is the same as how many pennies?

Simple Solutions© Mathematics

Level 2, 1st semester

1.	2.	3.
4.	5.	6.
7.	8.	9.
10.	11.	12.
13.	14.	15.

Simple Solutions® Mathematics — Level 2, 1st semester

Lesson #6

1. Which number comes between *19* and *34*?

    16   37   10   25

2. $3 + 5 = ?$

3. Which number is greater, *61* or *36*?

4. A box of spaghetti weighs about one pound. Does a bowling ball weigh more or less than a pound?

5. $14 - 6 = ?$

6. Which numbers are missing?   20, 30, ____, 50, 60, ____

7. Draw a heart and color one half of it.

8. $4 + 3 = ?$

9. What time is it on this clock?

10. $5 + 6 + 3 = ?$

11. Write the number that shows $20 + 8$.

12. Name the shape.

13. $9 - 2 = ?$

14. On what day of the week is December 25?

15. How many days are in December?

    What month comes after December?

		December				
S	M	T	W	T	F	S
		1	2	3	4	5
6	7	8	9	10	11	12
13	14	15	16	17	18	19
20	21	22	23	24	25	26
27	28	29	30	31		

1.	2.	3.
4.	5.	6.
7.	8.	9.
10.	11.	12.
13.	14.	15.

Lesson #7

1. Write the time that is shown on the clock.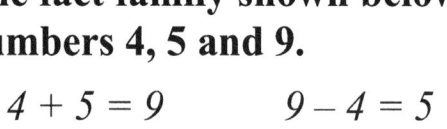

2. **A fact family is a set of related facts. The fact family shown below uses the numbers 4, 5 and 9.**

   $4 + 5 = 9$ $9 - 4 = 5$
   $5 + 4 = 9$ $9 - 5 = 4$

   Make a fact family using 3, 7 and 10.

3. Put a symbol in the circle. 86 ◯ 93

4. Does the + sign tell you to add or to subtract?

5. $12 - 9 = ?$

6. What shape comes next?

7. I have 3 tens and 2 ones. What number am I?

8. $6 + 7 = ?$

9. Write the name of the shape.

10. Write the number that comes right before 76.   ____, 76

11. How many parts out of 4 parts are shaded?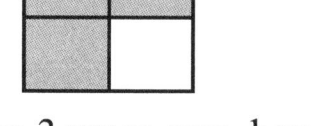

12. $11 - 4 = ?$

13. Terell has 9 toy cars. If his brother gives him 3 more cars, how many toy cars will he have?

14. $8 + 2 = ?$

15. How many red crayons are there?

    How many more red crayons than green ones are there?

Crayons	
Number	Color
16	Blue
10	Green
14	Red

1.	2.	3.
4.	5.	6.
7.	8.	9.
10.	11.	12.
13.	14.	15.

Lesson #8

1. What number comes in between *43* and *45*?   43 ____ 45

2. 8 + 4 = ?

3. Write the time shown on the clock.

4. Choose a sign to make this sentence true.
   54 ◯ 41   (>, <, =)

5. 9 − 2 = ?

6. The number *53* has ____ tens and ____ ones.

7. What number comes next?   15, 20, 25, ____

8. 3 + 7 = ?

9. **A *cone*** 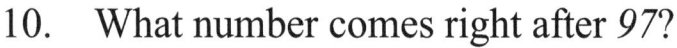 **is the same shape as a party hat.** Draw a cone in the box.

10. What number comes right after *97*?

11. Does a paper clip weigh more than or less than a pound?

12. Does the − sign tell you to add or to subtract?

13. Draw a square and shade three parts out of four parts.

14. 13 − 8 = ?

15. Write a fact family for the numbers 3, 9 and 12.

1.	2.	3.
4.	5.	6.
7.	8.	9.
10.	11.	12.
13.	14.	15.

Lesson #9

1. 8 − 6 = ?

2. What time is it?

3. What is the name of the shape?

4. 2 + 7 + 4 = ?

5. What fraction of the shape is shaded?

6. What number comes next in the list?   40, 50, 60, _____

7. 8 + 7 = ?

8. **There are 30 minutes in a half-hour.** Write *30 minutes = 1 half-hour* in the box.

9. How many sides does a triangle have?

10. There are 12 children on a soccer team and 7 of them are girls. How many boys are on the team?

11. Use a symbol to make this sentence true.   25 ◯ 16

12. Does a television weigh more than or less than a pound?

13. Write a fact family for 5, 6 and 11.

14. 8 tens and 2 ones are the same as what number?

15. How much money is the same as 3 dimes?

Simple Solutions© Mathematics — Level 2, 1st semester

1.	2.	3.
4.	5.	6.
7.	8.	9.
10.	11.	12.
13.	14.	15.

Lesson #10

1. Write a fact family for 8, 9 and 17.

2. 3 + 9 = ?

3. If you have three nickels, how much money do you have?

4. What number comes right before *63*?   _____, 63

5. Write the name of the shape?

6. 16 – 7 = ?

7. Choose a sign to make this true.   13 ◯ 51

8. The number *67* can also be written as _____ tens and _____ ones.

9. Write the time shown on the clock.

10. What part is shaded?

11. Yesterday, Ryan made 6 paper airplanes. He lost 2 of them. How many airplanes does he have left?

12. Which sign tells you to add?

13. Does it take about 1 minute or 1 hour to tie your shoes?

14. How many days are in a week?
    How many Wednesdays are in this month?

15. What month comes before May? After May?

May

S	M	T	W	T	F	S
			1	2	3	4
5	6	7	8	9	10	11
12	13	14	15	16	17	18
19	20	21	22	23	24	25
26	27	28	29	30	31	

1.	2.	3.
4.	5.	6.
7.	8.	9.
10.	11.	12.
13.	14.	15.

Lesson #11

1. 5 + 5 = ?

2. How much money is shown?

3. 12 − 8 = ?

4. Did Lin practice the piano for 1 minute or for 1 hour?

5. Write the name of the shape.

6. 37 ◯ 41

7. Write a fact family for 2, 7 and 9.

8. What number comes next in the list?   60, 70, 80, ____

9. How many days are in a week?

10. Write the time shown on the clock.

11. Greg got on base 6 times last week and 5 times this week. How many times did he get on base in the last two weeks?

12. Put these numbers in order from largest to smallest.   42, 18, 36

13. How many sides does a rectangle have?

14. **There are 12 months in a year.** Write *12 months = 1 year*.

15. The movie began at 4:00. It ended at 6:00. How long did the movie last?

1.	2.	3.
4.	5.	6.
7.	8.	9.
10.	11.	12.
13.	14.	15.

Lesson #12

1. The number *50* has 5 tens and _____ ones.

2. How many months are in a year?

3. 6 + 9 = ?

4. What is the time shown on the clock?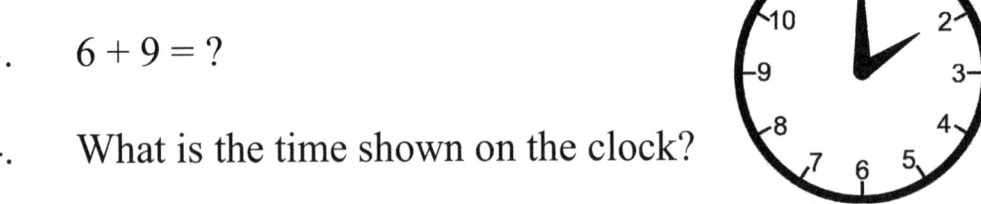

5. Does a piece of gum weigh more than or less than a pound?

6. What number comes before *61*? _____, 61

7. The number *25* has ____ tens and ____ ones.

8. Choose a sign to make this true. 16 ◯ 16

9. How much is shaded?

10. 18 − 9 = ?

11. Write a fact family for the numbers 4, 9 and 13.

12. Missy began with 16 cupcakes. She served 7 cupcakes for dinner. How many cupcakes did Missy have left?

13. 2 + 2 + 2 = ?

14. 70 + 9 is what number?

15. Christopher has 1 penny, 1 nickel, and 2 dimes. How much money does Christopher have?

1.	2.	3.
4.	5.	6.
7.	8.	9.
10.	11.	12.
13.	14.	15.

Lesson #13

1. **Numbers can be even or odd. Numbers are <u>even</u> if they end in 0, 2, 4, 6 or 8. Numbers are <u>odd</u> if they end in 1, 3, 5, 7 or 9. The number *36* is even because it ends in 6.** Is the number *47* even or odd? Why?

2. There are 13 fish in an aquarium. Four of the fish are goldfish. How many fish are <u>not</u> goldfish?

3. What time is it?

4. 6 + 8 = ?

5. Does a penny weigh more than or less than a pound?

6. Draw a rectangle and shade 1 part out of 3 parts.

7. 3 + 3 + 4 = ?

8. Another way to show *38* is with _____ tens and _____ ones.

9. 12 ◯ 19

10. 7 – 1 = ?

11. What number comes next in this list? 100, 200, 300, _____

12. Ten pennies equal one _____.

13. How many days are in one week?

14. Put in a sign to make this sentence true. 5 ◯ 6 = 11

15. At the zoo, 15 people are waiting in line to see the gorilla. If 8 people leave, how many people are still waiting to see the gorilla?

Simple Solutions© Mathematics — Level 2, 1st semester

1.	2.	3.
4.	5.	6.
7.	8.	9.
10.	11.	12.
13.	14.	15.

Lesson #14

1. Is *52* an even number or an odd number?

2. 4 + 7 = ?

3. What is the name of this shape?

4. 12 − 6 = ?

5. Write the number *sixty-four* in two other ways. (Hint: _____ tens and ____ ones or ____ + ____)

6. Count forward to *39*.   32, 33, ___, ___, ___, ___, ___, ___

7. If you have 2 dimes and 1 nickel, how much money do you have?

8. How many sides does a square have?

9. What time is shown on the clock?

10. Write the fact family for 4, 3 and 7.

11. Does it take about 10 minutes or 10 hours to eat your lunch?

12. Choose the sign to make this true.   37 ◯ 54

13. Write the number that comes right after *88*.   88, ____

14. There are 2 ears on 1 puppy. How many ears are on 3 puppies? Draw a picture to help you.

15. Which shape is divided into two equal parts? (Draw it in the box.)

Simple Solutions© Mathematics Level 2, 1st semester

1.	2.	3.
4.	5.	6.
7.	8.	9.
10.	11.	12.
13.	14.	15.

Lesson #15

1. 7 + 8 = ?

2. How many months are in a year?

3. 27 ◯ 16

4. Write the name of the shape.

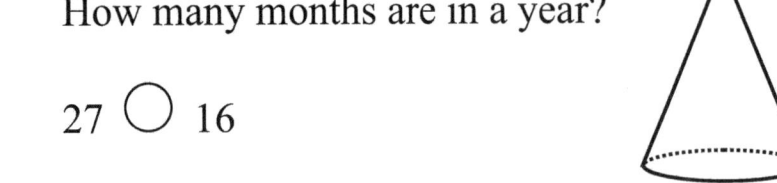

5. The number *seventy-five* has _____ tens and _____ ones.

6. Count backwards to *28*.   34, 33, 32, ____, ____, ____, ____

7. Write the sign that tells you to subtract.

8. 11 − 5 = ?

9. Put these numbers in order from smallest to largest.   74, 15, 26, 8

10. Write a fact family for 6, 7 and 13.

11. Does a television weigh more than a pound or less than a pound?

12. Write the time shown on the clock.

13. Marie has 21 stamps. If her sister gives her 6 more stamps, how many stamps will Marie have?

14. How many days are in 1 week?

15. How much money is shown below?

1.	2.	3.
4.	5.	6.
7.	8.	9.
10.	11.	12.
13.	14.	15.

Lesson #16

1. 10 − 3 = ?

2. Write the missing numbers. 35, ____, 45, 50, ____, ____, 65

3. Is the number *25* even or odd?

4. 34 ◯ 43

5. The number *92* has ____ tens and ____ ones.

6. 4 + 3 + 5 = ?

7. Which shape is divided into 4 equal parts? (Draw it in the box.)

8. How many months are in a year?

9. Two dimes and one nickel are the same as one _____.

10. Write the name of the shape.

11. **The answer in an addition problem is called the <u>sum</u>.** Write *sum* in the box.

12. Draw a rectangle and shade 1 part out of 3 parts. Another way to write this is $\frac{1}{3}$.

13. 32 + 5 = ?

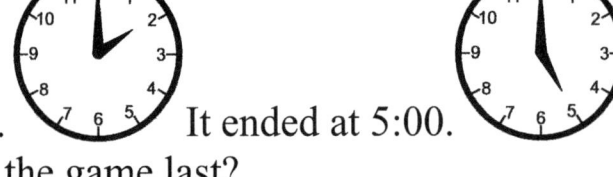

14. The game started at 2:00. It ended at 5:00. For how many hours did the game last?

15. **Use the tally table to answer the questions.**

    How many children like crackers?

    How many children like popcorn more than fruit?

Snacks We Like	Total
Popcorn	⁄⁄⁄⁄ ⁄⁄⁄⁄
Crackers	⁄⁄⁄⁄ /
Fruit	///

1.	2.	3.
4.	5.	6.
7.	8.	9.
10.	11.	12.
13.	14.	15.

Lesson #17

1. What is the answer to an addition problem called?

2. The number *354* has _____ hundreds, _____ tens and _____ ones.

3. 16 – 9 = ?

4. What time is shown on the clock?

5. Is the number *69* even or odd?

6. Fill in the missing numbers in the list. 20, ____, 40, ____, ____, 70

7. What comes next in the pattern? (Draw it in the box.)

8. 9 + 9 = ?

9. Does the eraser on a pencil weigh more than a pound or less than a pound?

10. Holly has 4 pennies and 4 dimes. How much money does she have?

11. 124 ◯ 59

12. What number comes between *38* and *40*?   38 ____ 40

13. How many parts out of 5 parts are shaded?

14. Write the fact family for 4, 8 and 12.

15. Read the thermometer. What is the temperature? ____°F

1.	2.	3.
4.	5.	6.
7.	8.	9.
10.	11.	12.
13.	14.	15.

Lesson #18

1. 3 + 3 + 2 = ?

2. What number comes before *57*? ____, 57

3. Put the numbers in order from greatest to least.   157, 298, 57

4. Write the time on the clock.

5. The number *314* has ___ hundreds, ___ tens and ___ ones.

6. 14 – 5 = ?

7. Write the next four even numbers.   0, ___, ___, ___, ___

8. 93 ◯ 41

9. Write *23* using words.

10. **A cylinder has the shape of a can.** Draw a cylinder in the box.

11. Write a fact family for 2, 9 and 11.

12. How much money is shown?

13. 8 + 6 = ?

14. The answer to an addition problem is called the _____.

15. There were 9 birds in a tree and then some more birds flew in. Now there are 16 birds in the tree. How many birds flew in? Make a picture to help you.

Simple Solutions© Mathematics — Level 2, 1st semester

1.	2.	3.
4.	5.	6.
7.	8.	9.
10.	11.	12.
13.	14.	15.

Lesson #19

1. What is the name of the shape?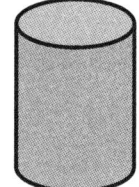

2. 7 + 7 = ?

3. Which digit is in the hundreds place in *426*?

4. 25 ◯ 50

5. Draw a square. Shade $\frac{1}{2}$ of it.

6. Do you go to bed at 8:30 am or at 8:30 pm?

7. **There are 365 days in a year.** Write *365 days = 1 year* in the box.

8. How many corners does a rectangle have?

9. Is the number *42* even or odd?

10. How many days are in a week?

11. Write the number *56* using words.

12. Write the fact family for 2, 9 and 11.

13. In an addition problem, the answer is called the _____.

14. Put these numbers in order from least to greatest. 96, 24, 155

15. Jennifer has 17 pieces of candy. If she gives 9 pieces to her friends, how many pieces of candy will she have left?

1.	2.	3.
4.	5.	6.
7.	8.	9.
10.	11.	12.
13.	14.	15.

Lesson #20

1. How many days are in a year?

2. Write the missing numbers in the list. 74, 76, ___, 80, ___, 84, ___

3. Does your desk weigh more than a pound or less than a pound?

4. Draw a triangle. How many sides does it have?

5. What is the time shown on the clock?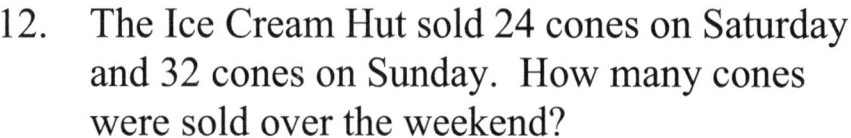

6. $9 - 4 = ?$

7. Is *64* an even or an odd number?

8. 33 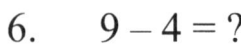 12

9. Which digit is in the tens place in *871*?

10. $13 + 16 = ?$

11. Write the standard number for *eighty-seven*.

12. The Ice Cream Hut sold 24 cones on Saturday and 32 cones on Sunday. How many cones were sold over the weekend?

13. $5 + 5 + 3 = ?$

14. Write the next four odd numbers. 1, ___ , ___, ___, ___

15. **Use the tally chart to answer the questions.**

    How many children chose red?

    How many more children chose blue than chose yellow?

Favorite Colors	
Blue	ՊՊՊ /
Red	ՊՊՊ ՊՊՊ
Yellow	////

1.	2.	3.
4.	5.	6.
7.	8.	9.
10.	11.	12.
13.	14.	15.

Lesson #21

1. Which digit is in the tens place in *68*?

2. 56 + 41 = ?

3. Write the next three even numbers. 0, ____, ____, ____

4. Five nickels are how much money?

5. 18 – 9 = ?

6. Put in the symbol that makes this true. 44 ◯ 76

7. Is *98* an even number or an odd number?

8. How many days are in one year?

9. Write the time shown on this clock.

10. 6 + 2 + 7 = ?

11. Does a piece of paper weigh less than or more than a pound?

12. Write a fact family for 5, 9 and 14.

13. What is the name of the shape?

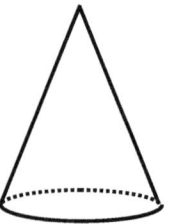

14. Write the number that comes next in the list. 12, 14, 16, ____

15. Ten students chose Math as their favorite subject. Twenty students picked Spelling as their favorite. How many more students like Spelling than Math?

1.	2.	3.
4.	5.	6.
7.	8.	9.
10.	11.	12.
13.	14.	15.

Lesson #22

1. Draw the shape that is divided into three equal parts.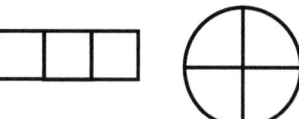

2. 34 + 23 = ?

3. Finish the pattern.

4. Write the standard number for *fifty-five*.

5. School began at 9:00 am. Lunch was at 12:00 pm. How many hours passed before lunch?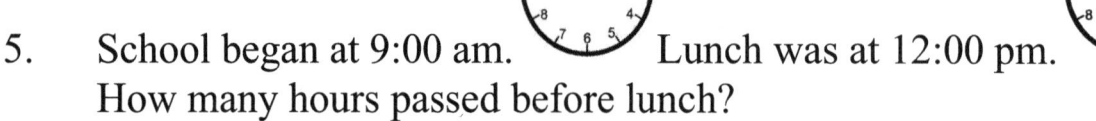

6. Would a watermelon weigh more than a pound or less than a pound?

7. Shade one part out of four parts. What fraction is shaded?

8. The number *72* can also be written as ___ tens and ___ ones.

9. Write the name of the shape.

10. 5 + 8 = ?

11. A half-hour is the same as _____ minutes.

12. Fill in the missing numbers.   72, ___, 76, 78, ___

13. Four nickels is the same amount of money as 2 _____.

14. 12 − 3 = ?

15. Mrs. Thomas made 16 cookies before lunch and 22 cookies after lunch. How many cookies did she make altogether?

1.	2.	3.
4.	5.	6.
7.	8.	9.
10.	11.	12.
13.	14.	15.

Lesson #23

1. How many minutes are in a half-hour?

2. $7 + 4 + 3 = ?$

3. Is *63* an even number or an odd number?

4. Fill in the symbol that makes this true. $18 \bigcirc 26$

5. Write the time shown on the clock.

6. How many months are in one year?

7. $12 - 6 = ?$

8. Dale worked two days last week. He earned $25.00 on Monday and $22.00 on Wednesday. How much did Dale earn last week?

9. Put these numbers in order from least to greatest. 26, 14, 39, 6

10. The answer to an addition problem is called the _____.

11. $66 + 13 = ?$

12. 3 tens and 4 ones equal what number?

13. Ten pennies equal one _____.

14. Which digit is in the hundreds place in *589*?

15. The second grade play is in 5 days. Today is January 15. On what date is the play?

January

S	M	T	W	T	F	S
			1	2	3	4
5	6	7	8	9	10	11
12	13	14	15	16	17	18
19	20	21	22	23	24	25
26	27	28	29	30	31	

1.	2.	3.
4.	5.	6.
7.	8.	9.
10.	11.	12.
13.	14.	15.

Lesson #24

1. How many corners does a triangle have?

2. $57 + 11 = ?$

3. **The answer to a subtraction problem is called the <u>difference</u>.** Write the word *difference* in the box.

4. There are how many days in a year?

5. Draw the shape that is divided into two equal parts.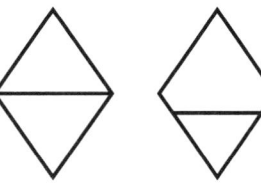

6. $4 + 9 = ?$

7. Write the name of the shape.

8. $43 \bigcirc 31$

9. Write the time on the clock.

10. $17 - 8 = ?$

11. How much money is shown?

12. Write the missing numbers in the list.
    55, 60, ____, 70, ____

13. Write the first four odd numbers.

14. Write the number *32* using words.

15. What temperature is shown on the thermometer?
    _____°F

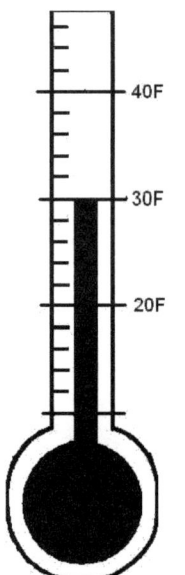

48

1.	2.	3.
4.	5.	6.
7.	8.	9.
10.	11.	12.
13.	14.	15.

Simple Solutions® Mathematics — Level 2, 1st semester

Lesson #25

1. 28 + 71 = ?

2. What is the answer to a subtraction problem called?

3. 75 ◯ 91

4. Write a fact family for 4, 6 and 10.

5. How many days are in a week?

6. Which digit is in the hundreds place in *238*?

7. What time is shown on the clock?

8. Does a bulldog weigh more than a pound or less than a pound?

9. 17 – 9 = ?

10. What number comes right before *67*?

11. Put in the sign that makes this true. 12 ◯ 5 = 7

12. The number *523* has ____ hundreds, ____ tens and ____ ones.

13. 5 + 4 = ?

14. Is the number *94* even or odd?

15. How many students chose a pig as their favorite farm animal?

    How many students voted altogether?

Favorite Farm Animals	
Cow	6
Horse	9
Pig	4

1.	2.	3.
4.	5.	6.
7.	8.	9.
10.	11.	12.
13.	14.	15.

Lesson #26

1. Write the number *127* using words.

2. 15 + 33 = ?

3. Does a ruler weigh more than a pound or less than a pound?

4. How many corners does a square have?

5. Count by tens. 30, ____, 50, 60, ____, ____

6. How many months are in a year?

7. 8 − 2 = ?

8. The answer to an addition problem is called the _____.

9. What is the name of the shape?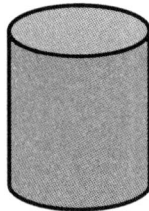

10. Write the sign that tells you to add.

11. One quarter = ____ ¢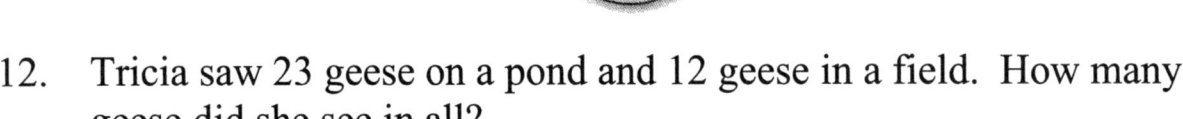

12. Tricia saw 23 geese on a pond and 12 geese in a field. How many geese did she see in all?

13. Finish the pattern.

14. Put the numbers in order from greatest to least. 28, 82, 38, 18, 33

15. Which shape is divided into two equal parts? (Draw it in the box.)

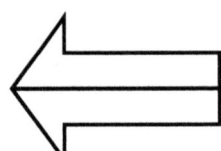 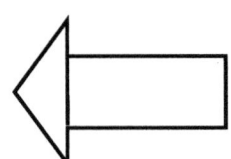

1.	2.	3.
4.	5.	6.
7.	8.	9.
10.	11.	12.
13.	14.	15.

Lesson #27

1. 31 + 57 = ?

2. What fraction is shaded?

3. Which digit is in the ones place in *438*?

4. Write the fact family for 6, 9 and 15.

5. Draw the shapes that have 4 sides.

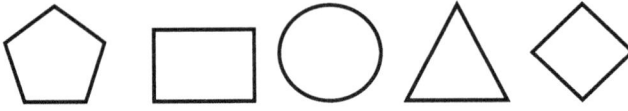

6. 15 – 7 = ?

7. Count by twos.   18, 20, ____, 24, ____, ____, 30

8. 18 ◯ 23

9. Write the first three even numbers.

10. 3 + 9 = ?

11. What is the time on the clock?

12. The number *83* has ____ tens and ____ ones.

13. How much money is this?

14. Put the numbers in order from least to greatest.   26, 13, 49, 37, 5

15. Jeff had 36 baseball cards. He gave 17 baseball cards to his brother. How many cards does he have left?

1.	2.	3.
4.	5.	6.
7.	8.	9.
10.	11.	12.
13.	14.	15.

Simple Solutions® Mathematics **Level 2, 1st semester**

Lesson #28

1. What number comes right after *73*?

2. Which digit is in the hundreds place in *136*?

3. 42 + 42 = ?

4. Which shape is $\frac{1}{3}$ shaded? Draw it in the box.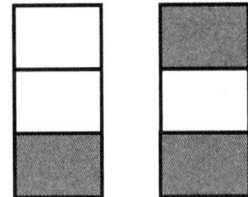

5. What is the name of the shape?

6. Write the standard number for *forty-two*.

7. 6 + 4 + 4 = ?

8. Twenty pennies equal how many dimes?

9. 33 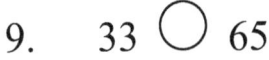 65

10. Write the fact family for 2, 8 and 10.

11. Does a bicycle weigh more than a pound or less than a pound?

12. Of the 48 dogs in the dog show, 16 were small dogs. How many large dogs were in the dog show?

13. The answer to a subtraction problem is called the _____.

14. There are _____ minutes in a half-hour.

15. 57 − 23 = ?

1.	2.	3.
4.	5.	6.
7.	8.	9.
10.	11.	12.
13.	14.	15.

Simple Solutions® Mathematics   Level 2, 1st semester

Lesson #29

1. Write the time shown on the clock.

2. How many days are in a year?

3. 19 ◯ 10

4. Does it take about 5 minutes or 5 hours to make your bed?

5. Fill in the missing numbers in the list. 55, ___, 65, ___, 75, ___

6. Over the summer Mark read 8 more books than James read. James read 5 more books than Jennifer did. Jennifer read 11 books. How many books did Mark read?

7. How much money is shown?

8. Put in a sign to make this true. 8 ◯ 5 = 13

9. 25 + 13 = ?

10. 6 tens and 4 ones are the same as what number?

11. 76 − 35 = ?

12. Numbers that end in 1, 3, 5, 7 or 9 are _____ numbers.

13. Draw the shape that has two equal parts.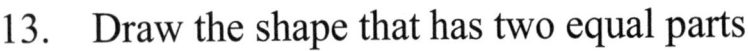

14. Put the numbers in order from least to greatest. 126, 38, 205, 79

15. How many sodas were sold?

    What was the total number of drinks sold at the stand?

Drink Stand	
Drink	Number Sold
soda	//
punch	ℍ
lemonade	ℍ ///
juice	////

58

1.	2.	3.
4.	5.	6.
7.	8.	9.
10.	11.	12.
13.	14.	15.

Lesson #30

1. 3 + 7 = ?

2. What is this coin worth?

3. How do you know if a number is even?

4. Write a fact family for 7, 8 and 15.

5. **There are 60 minutes in an hour.** Write *60 minutes = 1 hour* in the box.

6. 23 + 53 = ?

7. Which of the shapes is $\frac{1}{2}$ shaded? Draw it.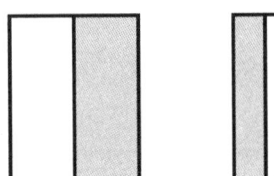

8. Six hundreds, 4 tens, and 7 ones equal what number?

9. 16 – 9 = ?

10. What is the time on the clock?

11. 68 – 22 = ?

12. Fill in the missing numbers. 49, 50, 51, _____, _____, _____, _____

13. 126 ◯ 243

14. The sum is the answer to a(n) _____ problem.

15. This morning there were 15 ducks on the pond. Some more ducks came and now there are 20 ducks on the pond. How many more ducks came to the pond?

Simple Solutions© Mathematics — Level 2, 1st semester

1.	2.	3.
4.	5.	6.
7.	8.	9.
10.	11.	12.
13.	14.	15.

Lesson #31

1. 6 + 5 = ?

2. Is the number *27* even or odd?

3. Count by tens. 60, 70, ____, ____, ____

4. There are _____ minutes in an hour.

5. 98 − 36 = ?

6. What is **value** of the coins? (How much money is this?)

7. How many months are in one year?

8. 257 ◯ 421

9. Write the time.

10. Put in a sign to make this number sentence true. 15 ◯ 9 = 6

11. Which digit is in the tens place in *485*?

12. Brad sold 5 boxes of candy on Monday. He sold 7 boxes of candy on Tuesday and 4 boxes on Wednesday. How many boxes of candy did Brad sell in all?

13. 29 + 50 = ?

14. Does a wooden baseball bat weigh more than or less than a pound?

15. Which shape has five sides? Draw it.

   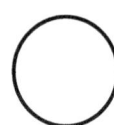

1.	2.	3.
4.	5.	6.
7.	8.	9.
10.	11.	12.
13.	14.	15.

Simple Solutions© Mathematics   Level 2, 1st semester

Lesson #32

1. Write the number *38* using words.

2. 816 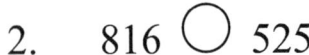 525

3. How much money is shown?

4. 31 + 18 = ?

5. What number comes right before *75*?

6. The number *516* has ___ hundreds, ___ tens and ___ ones.

7. Is *83* an even or an odd number?

8. Molly has to pay 7¢ a day for each overdue library book. She has two books to return. They are one day late. How much money will Molly have to pay in fines?

9. 77 − 43 = ?

10. The game started at 12:00 and ended at 12:30 . How long did the game last?

11. The answer to a subtraction problem is the _____.

12. How many days are in a year?

13. Does a postage stamp weigh more than or less than a pound?

14. 7 + 4 + 3 = ?

15. Write the temperature shown on the thermometer. ____°F

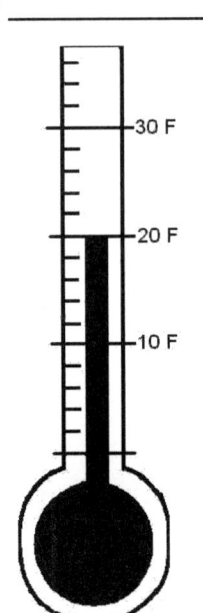

Simple Solutions© Mathematics Level 2, 1st semester

1.	2.	3.
4.	5.	6.
7.	8.	9.
10.	11.	12.
13.	14.	15.

Lesson #33

1. How many minutes are in a half-hour?

2. Draw a square. How many sides does it have?

3. Which digit is in the tens place in *827*?

4. 14 − 8 = ?

5. What is the time shown on the clock?

6. 312 ◯ 213

7. 2 + 8 + 4 = ?

8. Count by twos.   12, 14, ___, ____, 20

9. 65 + 33 = ?

10. The sum is the answer to a(n) _____ problem.

11. Write a fact family for 5, 7 and 12.

12. What is the value of the coins?

13. What number comes between *46* and *48*?   46 ____ 48

14. What is the name of the shape shown?

15. If 16 of the flowers in a garden are roses and 12 are tulips, how many more roses than tulips are in the garden?

1.	2.	3.
4.	5.	6.
7.	8.	9.
10.	11.	12.
13.	14.	15.

Lesson #34

1. Is *62* an even or an odd number?

2. 137 ◯ 317

3. Is the number *59* closer to 50 or to 60?

4. 54 − 31 = ?

5. How many minutes are in one hour?

6. Three nickels are equal to how many cents?

7. A year is the same as how many months?

8. Twenty-eight children chose pizza as their favorite food and sixteen children chose tacos. How many more children chose pizza than tacos?

9. Would a rubber band weigh more than or less than a pound?

10. 15 + 13 = ?

11. Write the standard number for *seventy-two*.

12. Which digit is in the hundreds place in *319*?

13. What part of the rectangle is shaded?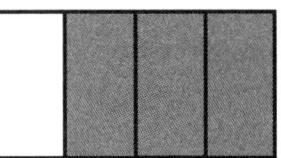

14. Put these numbers in order from greatest to least.   46, 21, 53, 17

15. Draw the four-sided shapes in the box.

1.	2.	3.
4.	5.	6.
7.	8.	9.
10.	11.	12.
13.	14.	15.

Lesson #35

1. Write the number for *two hundred thirty-six*.

2. Fill in the missing numbers. 200, 300, _____, _____, 600

3. 5 + 5 = ?

4. What time is it on this clock?

5. Draw a cone.

6. Write a fact family for 8, 9 and 17.

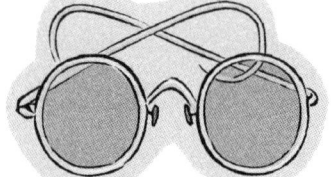

7. 68 − 21 = ?

8. The difference is the answer to a(n) _____ problem.

9. 396 ◯ 169

10. The number *806* has _____ hundreds, _____ tens and _____ ones.

11. A three-sided shape is called a(n) _____.

12. If you have ten dimes, how much money do you have?

13. 46 + 33 = ?

14. Fifteen children out of twenty-five children wear glasses. How many children do <u>not</u> wear glasses?

15. How many days are in April?

    April 10th is on what day of the week?

April

S	M	T	W	T	F	S
						1
2	3	4	5	6	7	8
9	10	11	12	13	14	15
16	17	18	19	20	21	22
23	24	25	26	27	28	29
30						

1.	2.	3.
4.	5.	6.
7.	8.	9.
10.	11.	12.
13.	14.	15.

Lesson #36

1. What number follows *86*?

2. Is the number *77* closer to 70 or to 80?

3. How many minutes are in an hour?

4. 9 + 2 = ?

5. What time is it?

6. 75 − 27 = ?

7. Write the next four even numbers.   0, ____, ____, ____, ____

8. Which shape is divided into two equal parts? Draw it.

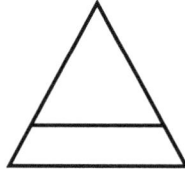 

9. Put in a sign to make this number sentence true.   18 ◯ 9 = 9

10. 35 + 26 = ?

11. Two nickels are the same as one _____.

12. Count backwards.   64, 63, ____, ____, ____, 59

13. 41 + 27 = ?

14. How many hundreds are in the number *415*?

15. There are five buttons on each of Pat's shirts. Pat has 4 shirts. How many buttons are there altogether? Draw a picture to help you.

1.	2.	3.
4.	5.	6.
7.	8.	9.
10.	11.	12.
13.	14.	15.

Lesson #37

1. Is *52* closer to 50 or to 60?

2. 36 + 45 = ?

3. What time is it?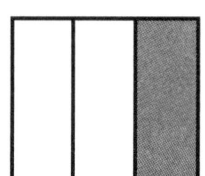

4. 10 − 4 = ?

5. Is the number *38* even or odd?

6. How many days are in a week? In two weeks?

7. 51 − 36 = ?

8. What part is shaded?

9. The answer to an addition problem is called the _____.

10. Does a bear weigh more than a pound or less than a pound?

11. Fill in the missing number.   46 ____ 48

12. Which digit is in the ones place in *76*?

13. How many minutes are in an hour?

14. Two quarters are how much money?

15. Juan and Jennifer cleaned up the litter near their house. They each collected 6 cans for recycling. How many cans did they collect?

1.	2.	3.
4.	5.	6.
7.	8.	9.
10.	11.	12.
13.	14.	15.

Lesson #38

1. Write the name of the shape.

2. Put in a sign to make this number sentence true.   16 ◯ 7 = 9

3. How many sides does this rhombus (diamond) have?

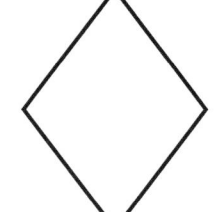

4. 63 + 22 = ?

5. Skip count by fives.   15, 20, ___, ___, ___

6. 14 + 26 = ?

7. Which digit is in the tens place in *324*?

8. Peter has 3 dimes and 1 nickel. How much money does he have?

9. 14 − 6 = ?

10. Write the number *fifty-three*.

11. **Rounding tells us about how many.** If there are 43 books on the shelf, are there about 40 or about 50 books? Round *43* to the nearest ten. (Hint: Is *43* closer to 40 or 50?)

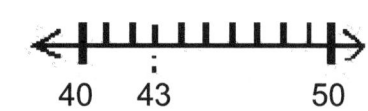

12. Write a fact family for 3, 9 and 12.

13. 57 ◯ 43

14. 3 + 7 + 4 = ?

15. **Use the bar graph for these questions.**
    How many hours did Lauren practice?

    Whose practice was shortest?

    Whose practice was longest?

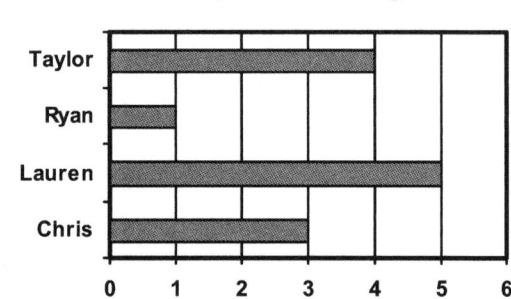

1.	2.	3.
4.	5.	6.
7.	8.	9.
10.	11.	12.
13.	14.	15.

Simple Solutions® Mathematics   Level 2, 1st semester

Lesson #39

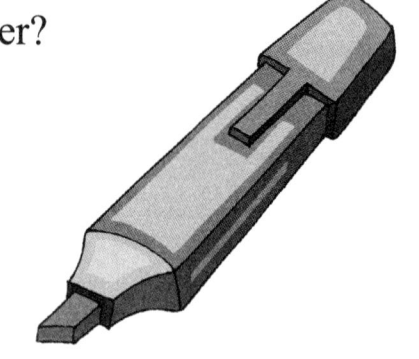

1. Is *35* an even number or an odd number?

2. 6 + 2 + 2 = ?

3. How many months are in a year?

4. 88 − 42 = ?

5. Count by twos.   18, 20, 22, ____, ____, ____

6. Write the time that is shown on the clock.

7. Which digit is in the hundreds place in *286*?

8. 25 + 36 = ?

9. What is the value of the coins?

10. Does a penny weigh more than or less than a pound?

11. How many minutes are in a half-hour?

12. Brad rode his bike 13 miles last week and 16 miles this week. How many miles did he ride altogether?

13. Put the numbers in order from least to greatest.   56, 87, 41, 30

14. Round *58* to the nearest ten. (Hint: Is it closer to 50 or to 60?)

15. Ellen has 13 markers. If she finds four more, how many markers will she have?

1.	2.	3.
4.	5.	6.
7.	8.	9.
10.	11.	12.
13.	14.	15.

Lesson #40

1. 86 ◯ 125

2. Is *33* an even or an odd number?

3. Which number comes right before *600*?

4. 5 + 8 = ?

5. The concert began at 2:00 and ended at 4:00. How long did the concert last?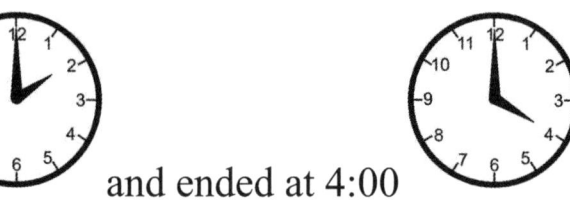

6. What is the name of the shape that is shown?

7. I have 3 tens and 5 ones. What number am I?

8. One nickel and two dimes are the same as one _____.

9. 45 + 45 = ?

10. Round *67* to the nearest ten. (Hint: Is it closer to 60 or to 70?)

11. Write the next number in the sequence. 20, 30, 40, ____

12. 75 − 38 = ?

13. Does it take about 15 seconds or 15 minutes to eat your breakfast?

14. How many days are in a year?

15. **Use the bar graph to help you.**

    How many people chose oranges as their favorite fruit?

    Which fruit got the fewest votes?

    Which got the most votes?

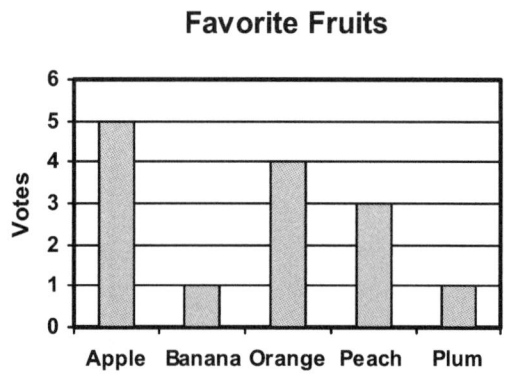

Favorite Fruits

1.	2.	3.
4.	5.	6.
7.	8.	9.
10.	11.	12.
13.	14.	15.

Lesson #41

1. $62 + 28 = ?$

2. Count by fives.   35, 40, ____, ____, ____

3. One dime and two quarters are equal to how much money?

4. Round *32* to the nearest ten.

5. How many days are in a year?

6. What number comes between 18 and 20?   18 ____ 20

7. $34 - 17 = ?$

8. Which digit is in the ones place in *51*?

9. Lynn is 3 years older than Cory. Cory is 5 years older than Sam. Sam is 10 years old. How old is Lynn?

10. 375 ◯ 815

11. Write a fact family for 5, 8 and 13.

12. Draw a triangle. How many sides does it have?

13. $3 + 5 + 5 = ?$

14. How many minutes are in an hour?

15. On which day of the week is January 1st?

    What is the date 2 weeks after January 10th?

    What month follows January?

January

S	M	T	W	T	F	S
		1	2	3	4	5
6	7	8	9	10	11	12
13	14	15	16	17	18	19
20	21	22	23	24	25	26
27	28	29	30	31		

1.	2.	3.
4.	5.	6.
7.	8.	9.
10.	11.	12.
13.	14.	15.

Lesson #42

1. How many months are in a year?

2. 77 + 23 = ?

3. Count by twos. 36, 38, ____, _____

4. What part is shaded?

5. How much money is shown?

6. Round *13* to the nearest ten.

7. Is the number *99* even or odd?

8. Write the number *37* using words.

9. What is the name of the shape?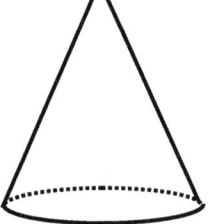

10. 596 ◯ 345

11. **This shape is called a cube.** 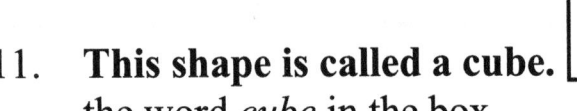 **It looks like an ice cube. Write the word *cube* in the box.**

12. 300 + 200 = ?

13. How much time has passed from one clock to the next?

14. I have 6 hundreds, 4 tens and 3 ones. What number am I?

15. Myron had 12 bagels. He ate 3 bagels and his friend, Shawn, ate 2 of them. How many bagels are left?

1.	2.	3.
4.	5.	6.
7.	8.	9.
10.	11.	12.
13.	14.	15.

Lesson #43

1. 4 + 3 + 6 = ?

2. Round *26* to the nearest ten.

3. Jason went for a walk at 2:00. He returned home from his walk at 3:30. How much time did Jason's walk take?

4. 47 + 29 = ?

5. Write a fact family for 2, 7 and 9.

6. What is the name of this shape?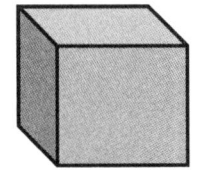

7. 125 + 342 = ?

8. Which digit is in the tens place in *206*?

9. What is the time shown on the clock?

10. The answer to a(n) _____ problem is called the difference.

11. Does an envelope weigh more than a pound or less than a pound?

12. 84 − 36 = ?

13. How much money is shown?

14. 288 ◯ 114

15. Use tally marks to show the number of combs.

1.	2.	3.
4.	5.	6.
7.	8.	9.
10.	11.	12.
13.	14.	15.

Lesson #44

1. 60 + 40 = ?

2. Write the standard number for *twenty-eight*.

3. Round *41* to the nearest ten.

4. I have 6 tens and 7 ones. What number am I?

5. 38¢ + 23¢ = ?

6. Draw a rectangle. How many sides does it have?

7. Write the time shown on the clock.

8. 73 − 26 = ?

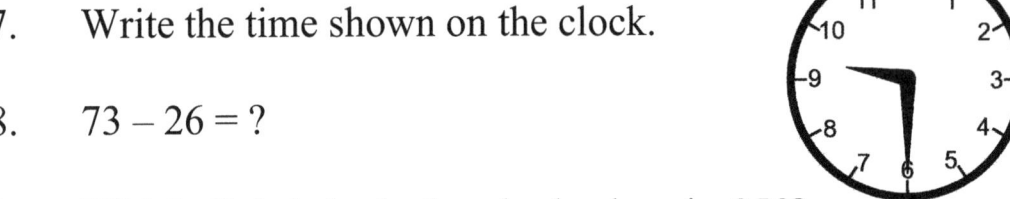

9. Which digit is in the hundreds place in *352*?

10. Count by tens.   60, 70, ____, ____, ____

11. 4 + 8 = ?

12. Write the name of the shape.

13. 63 ◯ 132

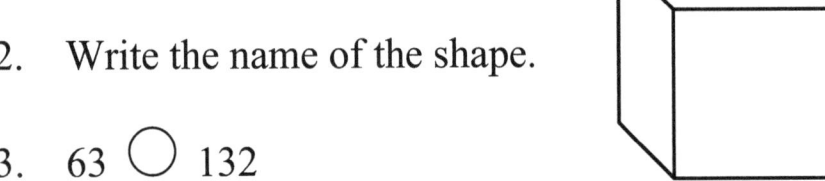

14. Does it take 10 minutes or 10 seconds to take a shower?

15. There were 47 toy trucks on a store shelf on Monday. By Wednesday, there were only 16 trucks left. How many trucks were sold from Monday to Wednesday?

1.	2.	3.
4.	5.	6.
7.	8.	9.
10.	11.	12.
13.	14.	15.

Lesson #45

1. Round *89* to the nearest ten.

2. What number comes right before *800*?

3. 64 + 28 = ?

4. Write a fact family for 4, 7 and 11.

5. 52 − 19 = ?

6. Is the number *137* even or odd?

7. The answer to an addition problem is called the _____.

8. How much time has passed from one clock to the next?

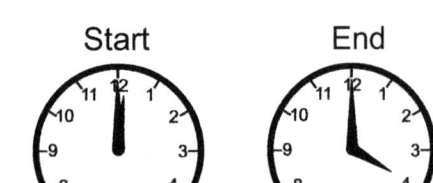

9. Write the number *83* using words.

10. 7 + 4 + 6 = ?

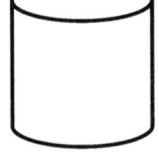

11. What is the name of the shape?

12. Which digit is in the tens place in *495*?

13. What is the value of the money?

14. 192 ◯ 503

15. **Use the graph to help you.**

    How many children like roses and sunflowers?

    How many more children like tulips than daisies?

Favorite Flowers

Simple Solutions® Mathematics — Level 2, 1st semester

1.	2.	3.
4.	5.	6.
7.	8.	9.
10.	11.	12.
13.	14.	15.

Lesson #46

1. $166 + 226 = ?$

2. Name the shape that is shown.

3. Which digit is in the ones place in *437*?

4. Count by twos.   44, 46, _____, _____

5. $50 - 28 = ?$

6. Does a napkin weigh more than a pound or less than a pound?

7. Round *18* to the nearest ten.

8. If you have 4 dimes, how much money do you have?

9. $33 + 54 = ?$

10. I have six hundreds, three tens and two ones. What number am I?

11. What time is it?

12. $86¢ + 14¢ = ?$

13. Draw a square.

14. **Two figures with the same size and shape** 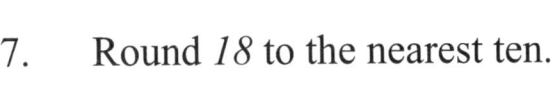 **are called congruent.** Draw two congruent squares in the box.

15. Yesterday, 49 planes took off from the Cleveland Airport. Today, 52 planes left the airport. How many planes left the airport in the last two days?

1.	2.	3.
4.	5.	6.
7.	8.	9.
10.	11.	12.
13.	14.	15.

Lesson #47

1. Draw 2 congruent circles.

2. 453 + 234 = ?

3. How much money is shown?

4. If Anna buys a doll for $18.25 and a stuffed animal for $9.55, how much money will she spend?

5. What time is shown on the clock?

6. 29 – 16 = ?

7. Would a large pumpkin weigh more than or less than a pound?

8. Round *39* to the nearest ten.

9. Write the next number in the sequence.   35, 45, _____

10. Which digit is in the hundreds place in *930*?

11. 4 + 3 = ?

12. How many months are in a year?

13. A three-sided figure is called a _____.

14. 15 – 7 = ?

15. Write the temperature shown on the thermometer. Would you go swimming at this temperature?

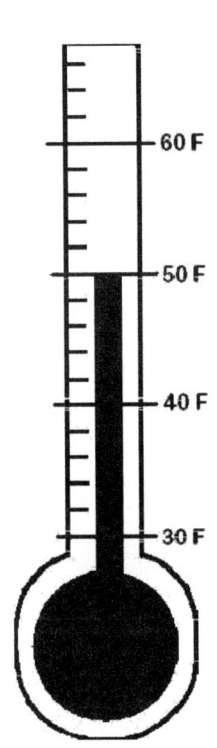

1.	2.	3.
4.	5.	6.
7.	8.	9.
10.	11.	12.
13.	14.	15.

Lesson #48

1. What month comes just before April?

2. 35 + 36 = ?

3. Would you measure the height of a tree in feet or in pounds?

4. Is the number *416* even or odd?

5. Draw two congruent rectangles.

6. Give the name of the shape.

7. 4 + 3 + 5 = ?

8. Write the standard number for *two hundred sixty-four*.

9. Round *42* to the nearest ten.

10. 53 − 18 = ?

11. If you have 5 nickels, how much money do you have?

12. At night, do you sleep about 8 minutes or 8 hours?

13. Write the symbol that shows subtraction.

14. 156 + 323 = ?

15. Mrs. Reilly drove 280 miles on Saturday and 142 miles on Sunday. How many miles did she drive in all?

Simple Solutions® Mathematics — Level 2, 1st semester

1.	2.	3.
4.	5.	6.
7.	8.	9.
10.	11.	12.
13.	14.	15.

Lesson #49

1. $349 + 137 = ?$

2. What is the time on the clock?

3. Two figures with the same size and shape are _____.

4. $96 - 38 = ?$

5. Write the number *65* using words.

6. $5 + 6 + 3 = ?$

7. Round *66* to the nearest ten.

8. Write the first four odd numbers.

9. What is the name of the shape?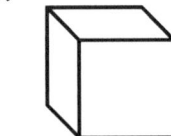

10. Two quarters and two dimes are equal to how much money?

11. I have seven tens and three ones. What number am I?

12. Which digit is in the ones place in *642*?

13. Which number comes between *63* and *85*?
          48   59   73   86

14. The pet shop has 16 puppies and 12 kittens. How many puppies and kittens are there altogether?

15. The answer to a subtraction problem is called the _____.

Simple Solutions© Mathematics — Level 2, 1st semester

1.	2.	3.
4.	5.	6.
7.	8.	9.
10.	11.	12.
13.	14.	15.

Lesson #50

1. Count by fives. 45, 50, 55, ____, ____, ____

2. 75 − 25 = ?

3. Put these numbers in order from greatest to least. 135, 79, 210, 56

4. Round *32* to the nearest ten.

5. 132 + 459 = ?

6. Draw two congruent rectangles.

7. 504 ◯ 396

8. What fraction is shaded?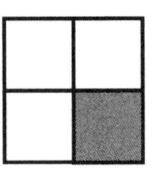

   $\frac{1}{2}$ $\frac{1}{3}$ $\frac{1}{4}$ $\frac{1}{5}$

9. 2 dimes and 1 nickel are the same amount of money as 1 _____.

10. There are how many minutes in a half-hour?

11. What number comes next in the list? 30, 40, 50, ____

12. There were 65 geese on a pond. Twenty-seven of the geese flew away. How many geese are left on the pond?

13. 876 − 154 = ?

14. How many days are in 2 weeks?

15. 36¢ + 29¢ = ?

1.	2.	3.
4.	5.	6.
7.	8.	9.
10.	11.	12.
13.	14.	15.

Lesson #51

1. Write the name of the shape.

2. $32 - 18 = ?$

3. Which shape is divided into two equal parts? Draw it.

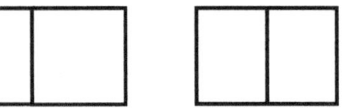

4. Would a bag of oranges weigh more than a pound or less than a pound?

5. On this clock, what time is it?

6. $226 + 342 = ?$

7. Round *89* to the nearest ten.

8. Which is longer, a month or a year?

9. 291 ◯ 86

10. Which digit is in the tens place in *475*?

11. $5 + 7 = ?$

12. How much money is shown?

13. Is *296* an even number or an odd number?

14. Which shape is not a rectangle? Draw it in the box.

15. How many students play the piano?

    How many more students play the piano than play the flute?

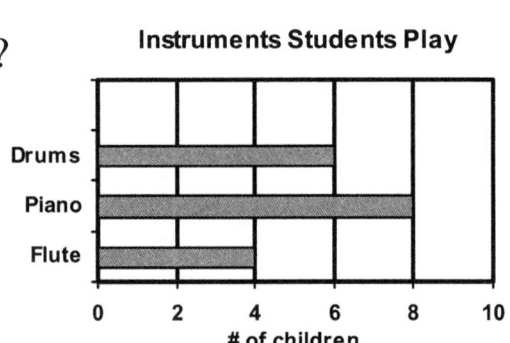

Instruments Students Play

Simple Solutions© Mathematics Level 2, 1st semester

1.	2.	3.
4.	5.	6.
7.	8.	9.
10.	11.	12.
13.	14.	15.

Lesson #52

1. 125 + 125 = ?

2. Count by twos.   56, 58, ____, ____

3. Round *27* to the nearest ten.

4. 98 ◯ 124

5. Write a fact family for 7, 8 and 15.

6. 56¢ + 27¢ = ?

7. What fraction is shaded?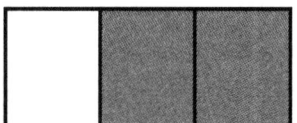

8. 83 − 35 = ?

9. Write the first four even numbers.

10. How much money is shown?

11. How many hours are in one day?

12. Two figures with the same size and the same shape are _____.

13. Which digit is in the hundreds place in *491*?

14. Write the standard number for *eighty-eight*.

15. Carlos had 15 goldfish. He won 4 more at the carnival. How many goldfish does Carlos have now?

1.	2.	3.
4.	5.	6.
7.	8.	9.
10.	11.	12.
13.	14.	15.

Lesson #53

1. Mia bought a piece of candy for 27¢. She paid the clerk 30¢. How much change did she get back?

2. 281 + 315 = ?

3. Draw 2 congruent ellipses (ovals).

4. 864 ◯ 746

5. Round *14* to the nearest ten.

6. 79 − 23 = ?

7. What fraction of the rectangle is shaded?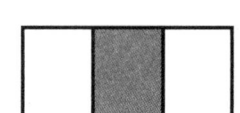

8. I have nine tens and four ones. What number am I?

9. How much time has passed?

10. Use tally marks to show the number *12*.

11. 8 + 8 + 4 = ?

12. How many minutes are in one hour?

13. Which number comes right before *816*?

14. Is the ride to school about 15 minutes or 15 seconds?

15. The answer to a(n) _____ problem is called the sum.

Simple Solutions® Mathematics — Level 2, 1st semester

1.	2.	3.
4.	5.	6.
7.	8.	9.
10.	11.	12.
13.	14.	15.

Lesson #54

1. Round *51* to the nearest ten.

2. 374 + 217 = ?

3. Four quarters are the same as one _____.

4. Write the next two numbers in the pattern. 55, 65, _____, _____

5. Draw two congruent squares.

6. 765 – 335 = ?

7. Order these numbers from least to greatest. 39, 11, 66, 43, 81

8. Write the time shown on the clock.

9. 855 ◯ 585

10. Draw a rectangle and shade $\frac{1}{2}$ of it.

11. Does a piece of gum weigh more than a pound or less than a pound?

12. Robert has three coins that add up to 25¢. What are his coins?

13. Which number comes between *156* and *172*?

    206   165   149   98

14. Count by twos. 28, 30, 32, ____, ____, ____

15. Kyle has 47 toy cars. Round the number of Kyle's cars to the nearest ten.

1.	2.	3.
4.	5.	6.
7.	8.	9.
10.	11.	12.
13.	14.	15.

Lesson #55

1. What time is shown on the clock?

2. Is *57* an even or an odd number?

3. Which digit is in the ones place in *370*?

4. Write the name of the shape.

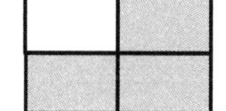

5. Round *69* to the nearest ten.

6. 653 ◯ 563

7. Write the number *124* using words.

8. What fraction of the rectangle is shaded?

9. 94 − 36 = ?

10. How much money is shown?

11. Draw 2 congruent diamonds.

12. 13 − 5 = ?

13. 357 + 238 = ?

14. **There are 12 inches in a foot. A ruler is a foot long.** Write *12 inches = 1 foot* in the box.

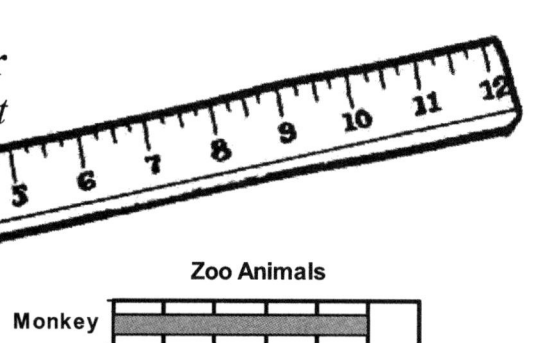

15. How many elephants are there?

    How many more monkeys than bears are at the zoo?

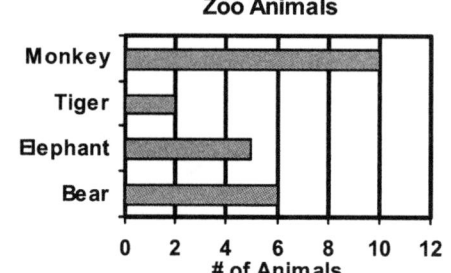

1.	2.	3.
4.	5.	6.
7.	8.	9.
10.	11.	12.
13.	14.	15.

Lesson #56

1. Is the number *355* even or odd?

2. How long is the line?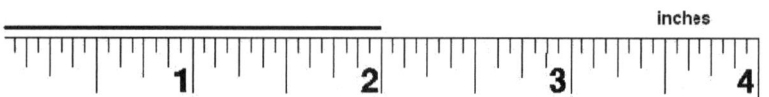

3. Write a fact family for 7, 9 and 16.

4. Round *12* to the nearest ten.

5. 227 + 567 = ?

6. What is the value of the coins?

7. 587 − 128 = ?

8. Two figures with the same size and shape are _____.

9. 88 $\bigcirc$ 15

10. Write the time shown on the clock.

11. 4 + 3 + 6 = ?

12. Order these numbers from greatest to least.   641, 298, 456, 192

13. Fran has twice as many pencils as Becky. Becky has 14 pencils. How many pencils does Fran have?

14. How many minutes are in a half-hour?

15. Write the day and the date that is 2 weeks after July 12th.

    What month comes after July?

July

S	M	T	W	T	F	S
					1	2
3	4	5	6	7	8	9
10	11	12	13	14	15	16
17	18	19	20	21	22	23
24	25	26	27	28	29	30
31						

Simple Solutions® Mathematics — Level 2, 1st semester

1.	2.	3.
4.	5.	6.
7.	8.	9.
10.	11.	12.
13.	14.	15.

Simple Solutions® Mathematics   Level 2, 1st semester

Lesson #57

1. Round *58* to the nearest ten.

2. Is the line 3 inches or 4 inches long?

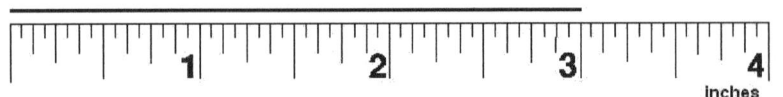

3. 329 ◯ 293

4. Write the first 5 odd numbers.

5. 86 + 77 = ?

6. I have 8 hundreds, 6 tens, and 2 ones. What number am I?

7. Which shape is <u>not</u> divided into 2 equal parts? Draw it.

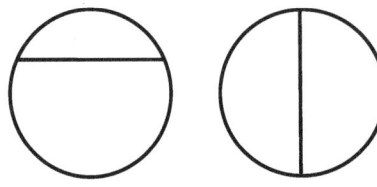

8. How many inches are in a foot?

9. Count by tens.   60, 70, ____, ____, ____

10. 452 – 139 = ?

11. Anne baked 3 dozen cupcakes. How many cupcakes did she bake?

12. Count the coins that are shown.

13. Josh bought a cookie for 20¢. He paid with a quarter. How much change did he get back?

14. Write the time shown on the clock.

15. Draw 2 congruent circles.

1.	2.	3.
4.	5.	6.
7.	8.	9.
10.	11.	12.
13.	14.	15.

Lesson #58

1. 173 + 415 = ?

2. Order these numbers from least to greatest. 485, 176, 321, 505

3. Round *47* to the nearest ten.

4. 314 ◯ 143

5. How much time has passed?

6. 90 − 26 = ?

7. How many inches are in a foot?

8. What month comes just before May?

9. What fraction is shaded?

10. Name the shape.

11. 7 + 5 = ?

12. Which digit is in the tens place in *613*?

13. Does a pretzel weigh more than a pound or less than a pound?

14. Which shape is not a four-sided shape?

15. What is the date of the 1st Thursday in October?

    Oct. 20th is on what day of the week?

    November 1st will be on what day of the week?

October

S	M	T	W	T	F	S
						1
2	3	4	5	6	7	8
9	10	11	12	13	14	15
16	17	18	19	20	21	22
23	24	25	26	27	28	29
30	31					

1.	2.	3.
4.	5.	6.
7.	8.	9.
10.	11.	12.
13.	14.	15.

Lesson #59

1. 266 + 326 = ?

2. 196 ◯ 205

3. What is the value of the coins?

4. Is a hammer about 11 inches or 11 feet long?

5. Write the next number in the pattern. 57, 59, 61, _____

6. 87 − 24 = ?

7. Is the number *398* even or odd?

8. How many months are in a year?

9. Round *72* to the nearest ten.

10. Write the number *214* using words.

11. What fraction of the figure is shaded?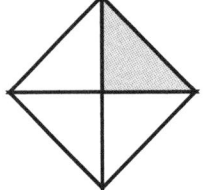

12. How long is this pencil? Give your answer in inches.

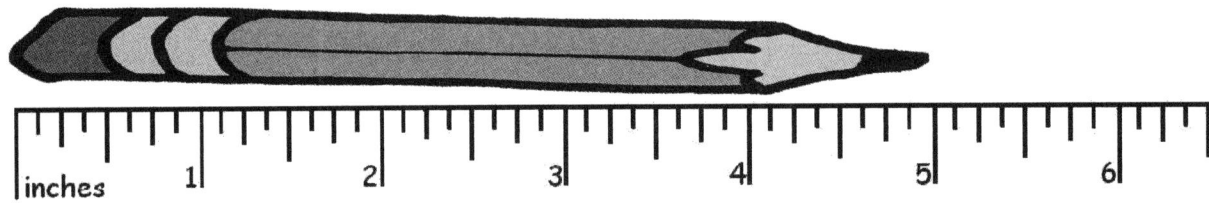

13. According to this clock, what time is it?

14. Use tally marks to show the number *18*.

15. Amanda bought a candy bar for 43¢. She paid 50¢. How much change did she get?

1.	2.	3.
4.	5.	6.
7.	8.	9.
10.	11.	12.
13.	14.	15.

Lesson #60

1. 933 + 57 = ?

2. Write the time shown on the clock.

3. Round *24* to the nearest ten.

4. 713 ◯ 45

5. Is your hand about 6 inches or 6 feet long?

6. Which digit is in the ones place in *92*?

7. 867 − 245 = ?

8. Name the shape.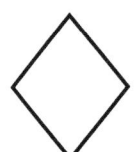

9. Mrs. Evans bought a purse that costs $35. She gave the clerk $50. How much change should she get back?

10. What fraction is shaded?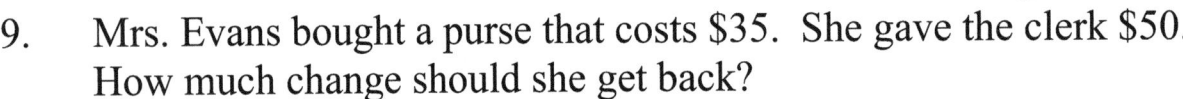

11. Two figures with the same size and shape are _____.

12. How many quarters are in a dollar?

13. Put these numbers in order from greatest to least.

    205   463   875   181

14. Does a peanut weigh less than or more than a pound?

15. What is the date of the 1st Friday in August?

    On what day of the week is August 19th?

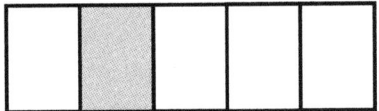

	August					
S	M	T	W	T	F	S
	1	2	3	4	5	6
7	8	9	10	11	12	13
14	15	16	17	18	19	20
21	22	23	24	25	26	27
28	29	30	31			

1.	2.	3.
4.	5.	6.
7.	8.	9.
10.	11.	12.
13.	14.	15.

Lesson #61

1. 37 + 21 + 13 = ?

2. What time is it?

3. 516 ◯ 615

4. Order these numbers from least to greatest. 507, 96, 321, 155

5. 590 − 235 = ?

6. How many inches are in a foot?

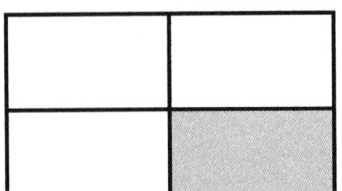

7. What fraction is not shaded?

8. David is 4 years younger than Tom. Tom is 3 years older than Vince. Vince is 8 years old. How old is David?

9. Write the name of this shape.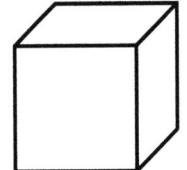

10. Which digit is in the tens place in *978*?

11. 435 + 227 = ?

12. What number comes just before *700*?

13. Three quarters equal how much money?

14. Does a box of books weigh more than or less than a pound?

15. What is the date on the second Wednesday in April?

    May 1st will be on what day of the week?

April

S	M	T	W	T	F	S
1	2	3	4	5	6	7
8	9	10	11	12	13	14
15	16	17	18	19	20	21
22	23	24	25	26	27	28
29	30					

1.	2.	3.
4.	5.	6.
7.	8.	9.
10.	11.	12.
13.	14.	15.

Lesson #62

1. How many days are in two weeks?

2. $58 + 39 = ?$

3. How much time has passed?

4. Round 88 to the nearest ten.

5. Which is longer, 2 weeks or one month?

6. $69¢ + 19¢ = ?$

7. I have 9 tens and 6 ones. What number am I?

8. 893 ◯ 983

9. Write the first five even numbers.

10. $342 + 329 = ?$

11. Two figures with the same size and shape are _____.

12. Write a fact family for 4, 9 and 13.

13. At the aquarium, there are 16 whales in one tank and 24 in the other tank. How many whales are there in all?

14. Draw two congruent rectangles.

15. How many more robins than sparrows were in the park?

    How many birds were in the park altogether?

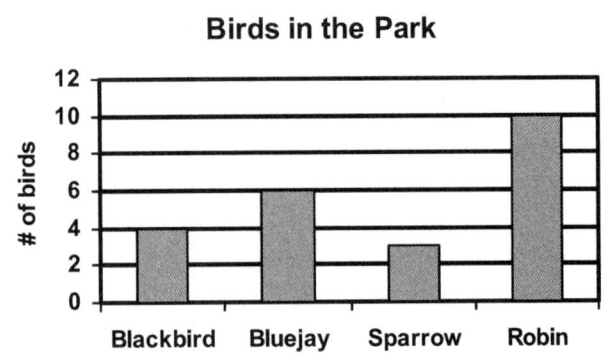

1.	2.	3.
4.	5.	6.
7.	8.	9.
10.	11.	12.
13.	14.	15.

Lesson #63

1. How many inches long is the line?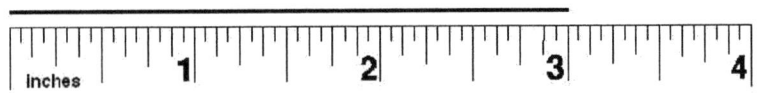

2. 42 + 15 + 24 = ?

3. Round *63* to the nearest ten.

4. Jared bought a baseball card for $1.50. He paid the clerk $2.00. How much change did he receive?

5. How many days are in two years?

6. Write the missing numbers.   42, 44, ____, 48, ____, 52

7. 318 ◯ 308

8. Is *418* an even or an odd number?

9. Which digit is in the hundreds place in *138*?

10. 530 − 119 = ?

11. What is the value of the coins?

12. Count by fives.   30, 35, ____, 45, ____, 55

13. What fraction is shaded?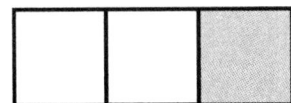

14. 408 + 264 = ?

15. Name the shape.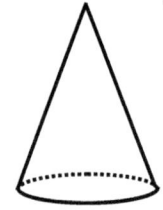

Simple Solutions® Mathematics — Level 2, 1st semester

1.	2.	3.
4.	5.	6.
7.	8.	9.
10.	11.	12.
13.	14.	15.

Lesson #64

1. Which is greater, 3 quarters or 10 dimes?

2. **This shape is called a pyramid.**
   Write *pyramid* in the box.

3. 653 − 239 = ?

4. Round 77 to the nearest ten.

5. Is a cell phone about 5 inches or 5 feet long?

6. What time is shown on the clock?

7. 448 + 217 = ?

8. What month comes after January?

9. A number has 6 hundreds, 7 tens and 5 ones. What is the number?

10. How many months are in a year?

11. 414 ◯ 404

12. Order these numbers from greatest to least. 815, 697, 904, 588

13. Mr. Martin used 2 feet of string for one project, 3 feet for another project and 4 feet for a third project. If he began with 16 feet of string, how much string was left?

14. Which shape is divided exactly in half? Draw it in the box.

15. What temperature is shown on the thermometer?

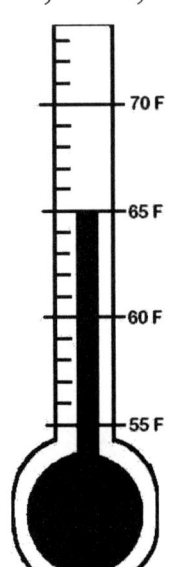

Simple Solutions® Mathematics — Level 2, 1st semester

1.	2.	3.
4.	5.	6.
7.	8.	9.
10.	11.	12.
13.	14.	15.

Lesson #65

1. Round *91* to the nearest ten.

2. Write the name of the shape.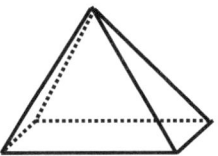

3. How many minutes are in an hour?

4. Which digit is in the ones place in *75*?

5. 463 + 428 = ?

6. What time is it?

7. Draw two congruent circles.

8. 318 ◯ 98

9. How many inches are in a foot?

10. 730 − 516 = ?

11. What number follows *699*?

12. Draw a square. Shade 1 part out of 4.

13. Which number comes next in the pattern?   63, 65, 67, _____

14. Kelvon has 63 baseball cards and 27 football cards. How many more baseball cards than football cards does he have?

15. What amount of money is shown?

1.	2.	3.
4.	5.	6.
7.	8.	9.
10.	11.	12.
13.	14.	15.

Simple Solutions® Mathematics   Level 2, 1st semester

Lesson #66

1. 28 + 14 + 33 = ?

2. What time is shown on the clock?

3. Round *88* to the nearest ten.

4. 649 − 125 = ?

5. Write the number *329* using words.

6. Write the name of the shape.

7. How long is the line?

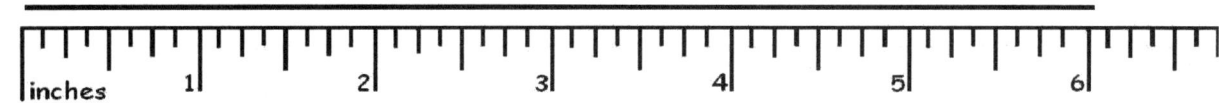

8. 329 + 464 = ?

9. How much money is shown?

10. Does one paper towel weigh more than or less than a pound?

11. 306 ◯ 74

12. Draw two congruent rectangles.

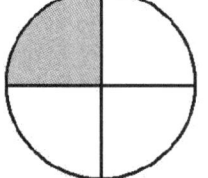

13. What fraction of the circle is shaded?

14. The answer to a subtraction problem is the _____.

15. Which subject is the favorite?

    How many more students like Math than English?

    How many students like Reading the best?

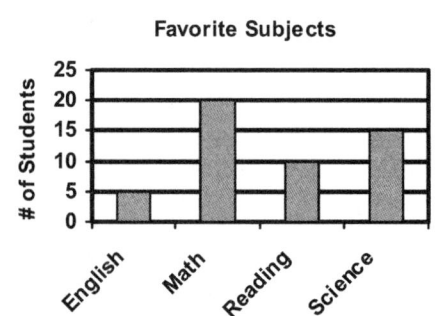

132

Simple Solutions© Mathematics — Level 2, 1st semester

1.	2.	3.
4.	5.	6.
7.	8.	9.
10.	11.	12.
13.	14.	15.

Simple Solutions® Mathematics Level 2, 1st semester

Lesson #67

1. Count by twos. 68, 70, ____, 74, ____, 78, ____

2. Does it take about 15 minutes or 15 hours to take a bath?

3. Round *63* to the nearest ten.

4. 507 + 384 = ?

5. Write the name of the shape.

6. How much money is ten nickels?

7. 80 − 25 = ?

8. A number has five tens and four ones. What is the number?

9. How many hours are in a day?

10. Write a fact family for 6, 9 and 15.

11. How much time has passed?

12. 6 + 5 + 7 = ?

13. Which digit is in the hundreds place in *317*?

14. Put these numbers in order from least to greatest.
    241   566   86   740

15. At a bake sale, cupcakes cost 50¢ each. Maria bought a cupcake and paid $1.00. How much change did she get back?

1.	2.	3.
4.	5.	6.
7.	8.	9.
10.	11.	12.
13.	14.	15.

Lesson #68

1. Is the number *530* even or odd?

2. 46 + 29 = ?

3. Round *75* to the nearest ten.

4. Which number comes between *41* and *83*?

    39   96   59   24

5. What time is it?

6. 233 ◯ 323

7. What month comes before August?

8. 973 − 544 = ?

9. What fraction of the rectangle is shaded?

10. Write the name of the shape.

11. The answer to an addition problem is called the _____.

12. Which is longer, 8 inches or 8 feet?

13. Write the standard number for *three hundred fifty-seven*.

14. Jessie rode her bicycle 12 miles on Monday and 9 miles on Tuesday. How many miles did Jessie ride in all?

15. Find the value of these coins.

Simple Solutions© Mathematics — Level 2, 1st semester

1.	2.	3.
4.	5.	6.
7.	8.	9.
10.	11.	12.
13.	14.	15.

Lesson #69

1. Is a stapler about 7 inches or 7 feet long?

2. Write the name of the shape.

3. 521 + 359 = ?

4. Round *68* to the nearest ten.

5. What time is shown on the clock?

6. 620 − 417 = ?

7. Which digit is in the ones place in *706*?

8. Two figures with the same size and shape are _____.

9. Draw a rectangle and shade 2 parts out of 3.

10. Jackie bought a whistle for 69¢. She gave the clerk 75¢. How much change did Jackie get?

11. 198 ◯ 301

12. How many inches are in a foot?

13. The answer to a(n) _____ problem is the difference.

14. 26 + 31 + 14 = ?

15. On what day of the week is November 15th?

    What is the date 3 weeks after November 4th?

    December 1st falls on what day of the week?

November

S	M	T	W	T	F	S
		1	2	3	4	5
6	7	8	9	10	11	12
13	14	15	16	17	18	19
20	21	22	23	24	25	26
27	28	29	30			

1.	2.	3.
4.	5.	6.
7.	8.	9.
10.	11.	12.
13.	14.	15.

Lesson #70

1. How much time has passed?

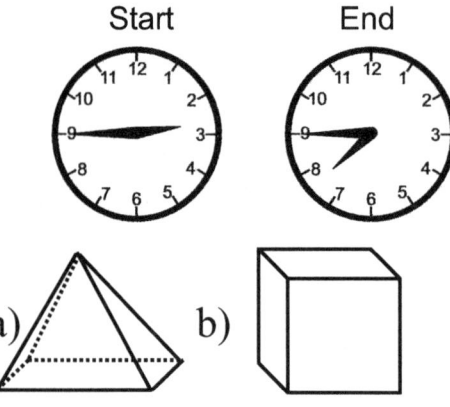

2. 76 − 39 = ?

3. Round *42* to the nearest ten.

4. Write the name of each shape. a) b)

5. 356 ◯ 216

6. Put these numbers in order from least to greatest.

   926   408   216   683

7. How much money is shown?

8. 637 + 228 = ?

9. Draw two congruent triangles.

10. What fraction of the circle is *not* shaded?

11. Count by fives.   35, 40, _____, 50, 55, _____

12. 63¢ + 37¢ = ?

13. Write a fact family for 8, 9 and 17.

14. I have six hundreds, zero tens and 4 ones. What number am I?

15. There are 27 students in Mrs. Kelly's third grade class and 29 students in Miss Gallo's third grade class. How many third graders are there altogether?

1.	2.	3.
4.	5.	6.
7.	8.	9.
10.	11.	12.
13.	14.	15.

Lesson #71

1. 14 − 9 = ?

2. What time is it?

3. 56 ◯ 65

4. Is a door 8 inches tall or 8 feet tall?

5. 535 + 259 = ?

6. Which figure is divided exactly in half? Draw it.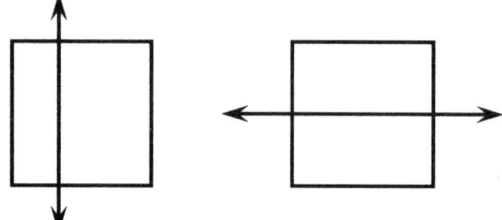

7. Round *15* to the nearest ten.

8. 128 + 325 = ?

9. Write the first four odd numbers.

10. 917 − 736 = ?

11. What is the name of this shape?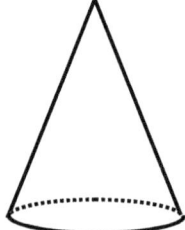

12. On Saturday, 324 children and 215 adults visited the zoo. Write the number sentence you would use to find out the total number of people who visited the zoo on Saturday.

13. Which is greater, 2 quarters or 6 dimes?

14. Write the number *689* using words.

15. How many days are in two years?

1.	2.	3.
4.	5.	6.
7.	8.	9.
10.	11.	12.
13.	14.	15.

Lesson #72

1. 193 − 79 = ?

2. Order the numbers from greatest to least.   518, 69, 247, 322

3. Round *22* to the nearest ten.

4. Gina is 4 years older than Mark. Mark is 6 years younger than Tim. Tim is 10 years old. How old is Gina?

5. 6 + 9 = ?

6. What number comes just before *614*?

7. $2.62 + $2.28 = ?

8. Write the time.

9. 1,000 ◯ 988

10. Which digit is in the tens place in *627*?

11. The answer to an addition problem is the _____.

12. How many inches are in a foot?

13. Find the value of these coins.

14. 595 + 327 = ?

15. Which two sports are the most popular?

    How many students voted for a sport?

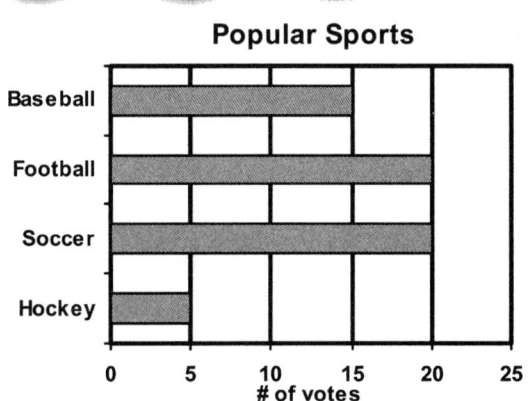

Popular Sports

1.	2.	3.
4.	5.	6.
7.	8.	9.
10.	11.	12.
13.	14.	15.

Level 2

Help Pages & "Who Knows?"

Help Pages

Vocabulary

Arithmetic Operations

Addition → When you combine numbers, you add. The sign "+" means add. The answer to an addition problem is called the *sum*. Example: When you combine 5 and 2, the sum is 7; 5 + 2 = 7.

Subtraction → When you take one number away from another, you subtract. The sign "-" means subtract. The answer to a subtraction problem is called the *difference*. Example: When you take 1 away from 5, the difference is 4; 5 - 1 = 4.

Multiplication → When you add a number to itself so many times, you multiply. The sign "x" means multiply. The answer to a multiplication problem is called the *product*. Example: When 5 is added to itself 3 times, the product is 15; 5 + 5 + 5 is the same as 3 x 5 = 15.

Division → When you share equally, you divide. The sign "÷" means divide. The answer to a division problem is called the *quotient*. Example: When 8 is shared equally between 2, the quotient is 4; 8 ÷ 2 = 4.

Geometry

Congruent — figures with the same shape and the same size.

Fraction — a part of a whole. Example: This box has 4 parts. 1 part is shaded. $\frac{1}{4}$

Line of Symmetry — a line along which a figure can be folded so that the two halves match exactly.

Geometry — Shapes and Solids

Cone —		Pyramid —	
Cube —		Rectangular Prism —	
Cylinder —		Rhombus (diamond) —	
Ellipse (oval) —		Sphere —	

148

Simple Solutions© Mathematics — Level 2, 1st semester

Help Pages

Vocabulary

Geometry — Polygons				
Number of Sides	Name	Number of Sides	Name
3 △	Triangle	4 ☐	Quadrilateral

Measurement — Relationships

Time	Distance
30 minutes = 1 half-hour	12 inches = 1 foot
60 minutes = 1 hour	**Volume**
365 days = 1 year	4 quarts = 1 gallon

Statistics

Mode — the number that occurs most often in a group of numbers. The mode is found by counting how many times each number occurs in the list. The number that occurs more than any other is the mode. Some groups of numbers have more than one mode.

Example: The mode of 77, ⑨③, 85, ⑨③, 77, 81, ⑨③, and 71 is **93**.
 (93 is the mode because it occurs more than the others.)

Place Value

Whole Numbers

1, 4 0 5

Thousands Hundreds Tens Ones

The number above is read: one thousand, four hundred five.

149

Simple Solutions© Mathematics — Level 2, 1st semester

Help Pages

Solved Examples

Whole Numbers (continued)

When we **round numbers**, we are estimating them. This means we focus on a particular place value, and decide if that digit is closer to the next higher number (round up) or to the next lower number (keep the same). It might be helpful to look at the place-value chart on page 149.

Example: Round 347 to the tens place.

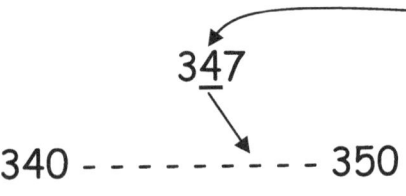

347 is closer to 350, so it is rounded to 350.

350

1. Identify the place that you want to round to.
2. What are the nearest "tens" on either side of the number? (340 and 350)
3. Which of these is 347 closer to?
4. This is the number you round to.

Here is another example of rounding whole numbers.

Examples: Round 83 to the nearest ten.

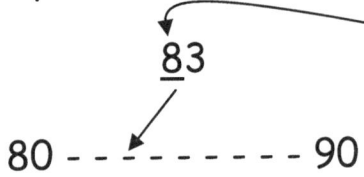

83 is closer to 80, so it is rounded to 80.

80

1. What is the rounding place?
2. What are the nearest "tens" on either side of the number? (80 and 90)
3. Which of these is 83 closer to?
4. This is the number you round to.

Help Pages

Solved Examples

Whole Numbers (continued)

There are **even numbers** and **odd numbers**. A number is <u>even</u> if it ends in 0, 2, 4, 6 or 8. A number is <u>odd</u> if it ends in 1, 3, 5, 7 or 9.

Examples: 46 is an even number because it ends in 6.

11 is an odd number because it ends in 1.

A **fact family** is a set of related facts using addition, subtraction, and the same three numbers.

Example: Write a fact family using 3, 4 and 7.

$3 + 4 = 7 \qquad 7 - 3 = 4$

$4 + 3 = 7 \qquad 7 - 4 = 3$

Numbers can be compared by saying one is **greater than** another or one is **less than** another.

The symbol ">" means *greater than*. The symbol "<" means *less than*.

Hint: The open part of the sign is near the bigger number.

Examples: 10 is less than 18. → 10 < 18

27 is greater than 13. → 27 > 13

Simple Solutions© Mathematics   Level 2, 1st semester

Help Pages

Solved Examples

Whole Numbers (continued)

When adding or subtracting whole numbers, first the numbers must be lined-up from the right. Starting with the ones place, add (or subtract) the numbers. When adding, if the answer has 2 digits, write the ones digit and regroup the tens digit. For subtraction, it may also be necessary to regroup first. Then, add (or subtract) the numbers in the tens place. Continue with the hundreds, etc.

Look at these examples of **addition**.

Examples: Find the sum of 314 and 12.   Add 648 and 236.

$$\begin{array}{r} 314 \\ +12 \\ \hline 326 \end{array}$$

1. Line up the numbers on the right.
2. Beginning with the ones place, add. Regroup if necessary.
3. Repeat with the tens place.
4. Continue this process with the hundreds place, etc.

$$\begin{array}{r} \overset{1}{6}48 \\ +236 \\ \hline 884 \end{array}$$

Use the following examples of **subtraction** to help you.

Example: Subtract 37 from 93.

$$\begin{array}{r} \overset{8}{\cancel{9}}\overset{13}{\cancel{3}} \\ -37 \\ \hline 56 \end{array}$$

1. Begin with the ones place. Check to see if you need to regroup. Since 7 is larger than 3, you must regroup to 8 tens and 13 ones.
2. Now look at the tens place. Check to see if you need to regroup. Since 3 is less than 8, you do not need to regroup.
3. Subtract each place value beginning with the ones.

Help Pages

Solved Examples

Whole Numbers (continued)

Example: Find the difference of 425 and 233.

$$\begin{array}{r} \overset{3\ 12}{\cancel{4}\cancel{2}5} \\ -2\ 3\ 3 \\ \hline 1\ 9\ 2 \end{array}$$

1. Begin with the ones place. Check to see if you need to regroup. Since 3 is less than 5, you do not need to regroup.
2. Now look at the tens place. Check to see if you need to regroup. Since 3 is larger than 2, you must regroup to 3 hundreds and 12 tens.
3. Now look at the hundreds place. Check to see if you need to regroup. Since 2 is less than 3, you are ready to subtract.
4. Subtract each place value beginning with the ones.

Sometimes when doing subtraction, you must **subtract from zero**. You will always need to regroup. Use the examples below to help you.

Example: Subtract 38 from 60.

$$\begin{array}{r} \overset{5\ 10}{\cancel{6}\cancel{0}} \\ -\ 3\ 8 \\ \hline 2\ 2 \end{array}$$

1. Begin with the ones place. Since 8 is less than 0, you must regroup.
2. Regroup to 5 tens and 10 ones.
3. Then, subtract each place value beginning with the ones.

Example: Find the difference between 500 and 261.

$$\begin{array}{r} \overset{4\ \overset{9}{\cancel{10}}\ 10}{\cancel{5}\cancel{0}\cancel{0}} \\ -2\ 6\ 1 \\ \hline 2\ 3\ 9 \end{array}$$

Help Pages

Solved Examples

Whole Numbers (continued)

Multiplication is a quicker way to add groups of numbers. The sign (×) for multiplication is read "times." The answer to a multiplication problem is called the product. Use the examples below to help you understand multiplication.

Example: 2 × 5 is read "two times five."

It means *2 groups of 5* or 5 + 5.

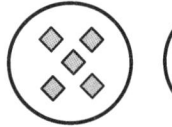 2 × 5 = 5 + 5 = **10**

The product of 2 × 5 is **10**.

Example: 5 × 4 is read "five times four."

It means *5 groups of 4* or 4 + 4 + 4 + 4 + 4.

 5 × 4 = 4 + 4 + 4 + 4 + 4 = **20**

The product of 5 × 4 is **20**.

It is very important that you memorize your **multiplication facts**. This table will help you, but only until you memorize them!

To use this table, choose a number in the top gray box and multiply it by a number in the left gray box. Follow both with your finger (down and across) until they meet. The number in that box is the product.

An example is shown for you:

2 × 5 = 10

×	0	1	2	5	10
0	0	0	0	0	0
1	0	1	2	5	10
2	0	2	4	10	20
5	0	5	10	25	50
10	0	10	20	50	100

Help Pages
Solved Examples

Whole Numbers (continued)

Division is the opposite of multiplication. The sign for division is ÷ and is read "divided by." The answer to a division problem is called the quotient.

Remember that multiplication is a way of adding groups to get their total. Think of division as the opposite of this. In division, you already know the total and the number in each group. You want to know how many groups there are. Follow the examples below.

Example: What is 9 ÷ 3?   (9 items divided into groups of 3)

  The total number is 9.

  Each group contains 3.

  How many groups are there?   There are 3 groups.

  9 ÷ 3 = **3**

Divide 10 by 2.   (10 items divided into groups of 2)

  The total number is 10.

  Each group contains 2.

  How many groups are there?   There are 5 groups.

  10 ÷ 2 = **5**

Fractions

A **fraction** is used to represent part of a whole. The top number in a fraction is the part. The bottom number in a fraction is the whole.

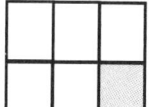

The whole rectangle has 6 sections.

Only 1 section is shaded.

This can be shown as the fraction $\frac{1}{6}$.

$\frac{1}{6}$ $\frac{\text{shaded part}}{\text{parts in the whole}}$

Simple Solutions© Mathematics — Level 2, 1st semester

Help Pages

Solved Examples

Time

The measure of how long something takes to happen is called **elapsed time**.

Example:

The movie began at 7:00 and ended at 9:00. How long did the movie last? (How much time passed between 7:00 and 9:00? There are **2 hours** between 7:00 and 9:00.

Example:

How many hours pass from the beginning of Spelling class until the end of Math class?

Spelling starts at 8:30. Math ends at 11:30. (How much time passes between 8:30 and 11:30?)

Class Schedule	
8:30 – 9:00	Spelling
9:00 – 10:00	Reading
10:00 – 11:30	Math
11:30 – 12:00	English

There are **3 hours** between 8:30 and 11:30.

Help Pages
"Who Knows?"

Sides in a triangle?..(3)

Sides in a square?...(4)

Days in a week? ....(7)

Months in a year? ....(12)

Days in a year? ....(365)

Inches in a foot?...(12)

Quarts in a gallon?...(4)

The number that is seen most
often in a set of numbers?........................... (mode)

Figures with the same size
and shape? .... (congruent)

Answer to an addition problem?.....................(sum)

Answer to a subtraction problem?....(difference)

Answer to a multiplication
problem?..(product)